essentials

Springer essentials

Springer essentials provide up-to-date knowledge in a concentrated form. They aim to deliver the essence of what counts as "state-of-the-art" in the current academic discussion or in practice. With their quick, uncomplicated and comprehensible information, *essentials* provide:

- an introduction to a current issue within your field of expertise
- an introduction to a new topic of interest
- an insight, in order to be able to join in the discussion on a particular topic

Available in electronic and printed format, the books present expert knowledge from Springer specialist authors in a compact form. They are particularly suitable for use as eBooks on tablet PCs, eBook readers and smartphones. *Springer essentials* form modules of knowledge from the areas economics, social sciences and humanities, technology and natural sciences, as well as from medicine, psychology and health professions, written by renowned Springer-authors across many disciplines.

Gerd Thomas Waldhauser

Neuropsychoanalysis

An Introduction to Neuroscience and Psychodynamic Therapy

Gerd Thomas Waldhauser
Department of Neuropsychology, Institute
of Cognitive Neuroscience, Faculty of Psychology
Ruhr University Bochum
Bochum, Germany

ISSN 2197-6708 ISSN 2197-6716 (electronic)
essentials
ISSN 2731-3107 ISSN 2731-3115 (electronic)
Springer essentials
ISBN 978-3-658-40890-9 ISBN 978-3-658-40891-6 (eBook)
https://doi.org/10.1007/978-3-658-40891-6

This book is a translation of the original German edition "Neuropsychoanalyse" by Waldhauser, Gerd Thomas., published by Springer Fachmedien Wiesbaden GmbH in 2021. The translation was done with the help of artificial intelligence (machine translation by the service DeepL.com). A subsequent human revision was done primarily in terms of content, so that the book will read stylistically differently from a conventional translation. Springer Nature works continuously to further the development of tools for the production of books and on the related technologies to support the authors.

© The Editor(s) (if applicable) and The Author(s), under exclusive license to Springer Fachmedien Wiesbaden GmbH, part of Springer Nature 2023

This work is subject to copyright. All rights are solely and exclusively licensed by the Publisher, whether the whole or part of the material is concerned, specifically the rights of translation, reprinting, reuse of illustrations, recitation, broadcasting, reproduction on microfilms or in any other physical way, and transmission or information storage and retrieval, electronic adaptation, computer software, or by similar or dissimilar methodology now known or hereafter developed.

The use of general descriptive names, registered names, trademarks, service marks, etc. in this publication does not imply, even in the absence of a specific statement, that such names are exempt from the relevant protective laws and regulations and therefore free for general use.

The publisher, the authors, and the editors are safe to assume that the advice and information in this book are believed to be true and accurate at the date of publication. Neither the publisher nor the authors or the editors give a warranty, expressed or implied, with respect to the material contained herein or for any errors or omissions that may have been made. The publisher remains neutral with regard to jurisdictional claims in published maps and institutional affiliations.

This Springer imprint is published by the registered company Springer Fachmedien Wiesbaden GmbH, part of Springer Nature.
The registered company address is: Abraham-Lincoln-Str. 46, 65189 Wiesbaden, Germany

Paper in this product is recyclable.

What You Can Find in This *essential*

- An insight into the neuroscientific basis of psychoanalytic concepts
- An overview of the history and methods of neuropsychoanalysis
- A discussion of the perspective and limitations of an exchange between neuroscience and psychoanalysis

Contents

1 **Introduction** 1
   1.1 What Is Neuropsychoanalysis? 2

2 **History and Problems of the Exchange Between Neuroscience and Psychoanalysis** 3
   2.1 Historical Overview 3
   2.2 Epistemological Considerations 5

3 **Neuropsychoanalytical Research** 9
   3.1 Operating Principles of the Mindbrain 9
       3.1.1 Drive and Emotion 10
       3.1.2 Learning Through Experience 12
       3.1.3 The Neural Basis of the Structural Model 14
   3.2 Memory and Defense 17
       3.2.1 Memory Systems and Memory Processes 18
       3.2.2 Repression and Defense Mechanisms 20
   3.3 Developmental and Social Psychological Perspectives 25
   3.4 Neuroscientific Therapy Research 28

4 **Perspectives and Limitations** 31

What You Learned from This *essential* 35

References 37

Introduction 1

Psychoanalysis as a science has had an increasingly difficult time at universities for some years now. At the same time, within psychoanalytic discourse, scientific methods and findings are partially received, but dismissed by some psychoanalysts as irrelevant to practice (Chap. 4). This book aims to counter the split between cognitive neuroscientific research and psychoanalytic practice using reasoned arguments, by citing empirical evidence, and by elaborating links between the fields of research.

The recognition of psychoanalysis as an applied therapeutic method requires scientific, empirical evidence for its effectiveness and for the manifold and complex psychoanalytic concepts. One can argue against this demand by saying that the scientific method cannot do justice to the individual and subjective processes in psychoanalytic therapy. However, this argumentation seems unrealistic in view of the distributional struggles in health care and research.

The now commonly held belief that psychoanalysis is "unscientific" is false. Many psychoanalytic concepts can be directly or indirectly substantiated or linked to findings in the affective, social, and cognitive neurosciences (see Chap. 3). Even if psychoanalysis lags behind other schools of therapy in terms of the number of scientific studies, it is debatable whether tracing some concepts of cognitive behavioral therapy to laboratory findings or animal models is more viable than, for example, the empirical basis for transference phenomena. If one follows the Nobel Prize winner Eric Kandel, psychoanalysis still represents the most comprehensive theory of the human mind in the twenty-first century (Kandel 1999). It should be just as much in the interest of neuroscience to take this body of theory seriously as it makes sense for psychoanalysts to engage with neuroscientific findings.

© The Author(s), under exclusive license to Springer Fachmedien Wiesbaden GmbH, part of Springer Nature 2023
G. T. Waldhauser, *Neuropsychoanalysis*, Springer essentials,
https://doi.org/10.1007/978-3-658-40891-6_1

1.1 What Is Neuropsychoanalysis?

In order to describe the dialogue between neuroscience and psychoanalysis comprehensively, the term "neuropsychoanalysis" was coined by probably the most important and influential representative of this research approach, Mark Solms (Kaplan-Solms and Solms 2000; Solms and Turnbull 2011). Although there had already been manifold exchanges between the two disciplines before, the creation of a separate term shows that, since the late 1990s, there has been a scientific community that is decidedly dedicated to the interdisciplinary study of neuropsychological mechanisms from a psychoanalytic point of view and vice versa. Neuropsychoanalysis sees itself as transcending and open to the various neuroscientific subdisciplines and psychoanalytic schools of thought. Thus, within the neuropsychoanalytic community there are self-psychological, Freudian, Lacanian, intersubjective, and ego-psychological approaches. Insights from affective, cognitive, social, and computational neuroscience are incorporated. Since the emergence of the term, several core assumptions of neuropsychoanalysis have emerged. These include, for example, the drive and affect theories of Panksepp (1998) or the predictive coding framework (Carhart-Harris and Friston 2010). However, these core assumptions are by no means to be understood as a central characterization of neuropsychoanalysis.

Neuropsychoanalytic research can be broadly divided into three subfields:

1. Clinical neuropsychology from a psychoanalytic perspective: Case studies from psychoanalytic work with neurological patients
2. Psychoanalytically oriented neuroscience: Experimental investigations of psychoanalytic concepts with neuroscientific methods
3. Neuroscientific therapy research: investigation of brain physiological changes after psychodynamic psychotherapy

In Chap. 3 of this book, I will present important core findings from these areas. As an experimental memory researcher, my focus is on the second subfield, with a main emphasis on the investigation of defense mechanisms and dynamic memory models. I will also include findings that did not originate with a decidedly neuropsychoanalytical objective, but are highly relevant to the topic.

History and Problems of the Exchange Between Neuroscience and Psychoanalysis

Neuropsychoanalysis sounds like an attempt to square the circle. Psychoanalysis and neuroscience could not be more different in many respects: "scenical comprehension", hermeneutics, radical subjectivity and non-replicability on the one hand, experimental methodology, quantifiability and objectifiability on the other. Nevertheless, for over 120 years there have been repeated attempts to unite the two sides. The following overview will sketch the history of this exchange. Subsequent epistemological considerations on the theory of science will outline the compromises that have to be made in order to enable research in this area.

2.1 Historical Overview

The desire to reconcile psychoanalysis and neuroscience can be traced back to the early days of psychoanalysis. Freud, significantly influenced by his training as a neurologist, elaborated in his posthumously published "Project for a Scientific Psychology" (1950 [1895]) his intention to find a basis for psychodynamic processes at the neurophysiological level. Not only did he follow the then-currently debated neuron theory, but he postulated different specialized nerve cell types interacting with each other in a dynamic way, via quantifiable excitation patterns and their inhibition. Moreover, from the outset Freud was concerned with how processes that significantly characterized his clinical considerations might be concretely realized in the human brain and nervous system. Freud did not pursue this approach further, however, as he found that the neurophysiological knowledge of his time was insufficient to represent the processes he had discovered. However, the wish that this would eventually succeed did not let him go. Thus, in "On Narcissism" (1914b, p. 78), he writes that one must "must recollect that all our provisional ideas in psychology will presumably some day be based on an organic substructure."

Scientific thinking and methodology were also not foreign to Freud's companions. In his diagnostic association experiments, Carl Gustav Jung used psychophysiological measurement methods and reaction time measurements (Jung and Ricklin 1906) to find empirical evidence for repression mechanisms. In experimental psychology, which was emerging at the same time, there were a large number of studies examining the influence of emotions on memory or attention that drew directly on psychoanalytic ideas (Rapaport 1971). There were even attempts to explain and formalize repression in terms of learning theory, thus reconciling psychoanalytic ideas with the behaviorist paradigm (Dollard and Miller 1950). These ideas laid the foundation for experimental, neuroscientific investigation of psychoanalytic concepts. With the discovery and development of electroencephalography (EEG), neuroscientific investigations of defense mechanisms were realized as early as the 1950s and some were highly published (Dixon and Lear 1963). Encouraged by the conceptual proximity of ego psychology and cognitive psychology, some research groups worked on an information-theoretical formalization of psychoanalytic theory (Erdelyi 1985; Koukkou and Lehmann 1998). From the 1960s onwards, however, the cross-connections between (cognitive) neuroscience and psychoanalysis diminished markedly, and psychodnamic thinking became at best a marginal phenomenon within academic psychology in the following decades.

The psychoanalytic community largely followed Freud's decision to initially shelve a neurophysiological justification of his theories. Psychoanalysis continued to develop its own hermeneutic methods and oriented itself more towards neighbouring disciplines in the cultural sciences and humanities than towards the natural sciences (Warsitz and Küchenhoff 2015). Some psychoanalytic researchers resumed the "other thread" of psychoanalysis that Freud had set aside. Lorenzer, for example, explicitly extended the object of research of psychoanalysis beyond the linguistic and cultural to the body and thus also the neuronal (Lorenzer 2002).

However, it was to take until the nineties of the twentieth century for the concept and notion of "neuropsychoanalysis" to develop. This was facilitated by the immense methodological progress in the neurosciences. With functional magnetic resonance imaging, a non-invasive imaging technique was established that made it possible to map correlates of mental processes by measuring blood oxygen levels in the human brain. In addition, the network character of brain activity could be confirmed by so-called "time-frequency methods" in the EEG and measured in real time. The knowledge of neurotransmitters, complex systems and cognition and, last but not least, the availability of high computing power for the analysis of large amounts of data, enabled new insights into the human brain. A neuroscientific explanation of even complex mental phenomena, such as the emergence of consciousness, seemed within reach. So it was only natural to launch another attempt to locate even the unwieldy and elusive ideas of psychoanalysis in the brain (Kaplan-Solms and Solms 2000; Solms 2000; Koukkou et al. 1998). However, the next section will demonstrate that this is by no means a simple undertaking.

2.2 Epistemological Considerations

Any attempt to relate psychological phenomena to physiological processes in the brain must confront two important philosophical questions:

1. How are mind and body related?
2. Which methods are appropriate to study the object of investigation?

Both questions are difficult to answer in themselves and are the cause of far-reaching debates. They are also intertwined: different methods are required depending on what exactly is identified as the object of investigation. The challenge of finding answers to both questions arises for neuropsychoanalysis in particular, since it conceives of the mind as an essentially subjective phenomenon, but the brain as a material thing that can be studied from a third-person perspective. It claims both to understand the subjective experience of the individual and his unconscious desires, and to make generally valid statements about the psyche that can be grounded in neuroscience.

Is the Mind Nothing but the Brain?

How can conscious, subjective phenomenal perceptions and sensations emerge from the physical matter of our brains? The marriage of psychoanalysis and neuroscience is particularly confronted with this "hard problem" of consciousness because psychoanalysis places a special value on subjective experience (Chalmers 1995). In other disciplines, the hard questions about the quality of experience can be bypassed or broken down into many "easy problems" that can be studied at the functional or physical level of behavior (Dennett 2018). However, this does not seem to apply to psychoanalysis, where the belief is that subjective experience matters in its wholeness (Lorenzer 2002). So what stance can one take if one considers both subjective experience and the findings of neuroscience to be essential to the advancement of knowledge?

Representatives of materialistic positions on the brain-mind relationship assume that, at least in principle, everything mental can be fully explained by material processes. From this point of view, the endeavour to deal with mental processes does not bring any progress in knowledge, provided that one has the necessary methods to investigate and understand the brain completely. The radical opposite position of idealism and also some representatives of functionalism, on the other hand, deny any relevance of the material world for the understanding of the human mind, and dualism assumes that mind and body are different entities. All of these positions challenge either one or the other side of the neuropsychoanalytic endeavor. Solms and colleagues propose the position of two-aspect monism (e.g., Solms 2019; Yovell et al. 2015): This postulates that the brain and mind are inextricably linked and can only be considered as a "mindbrain" unit. The mindbrain has two distinct aspects, each of which can only be captured via a particular approach. The "mind" aspect is ultimately only accessible through subjective experience, while the "brain" aspect cannot be examined from an inner world perspective, but only from an outer world perspective. It is now no longer necessary to ask how inner experience arises from the brain, but to assume that the two objects of investigation represent different sides of the same coin. The position is not unproblematic for a variety of reasons, but it does allow the work of those concerned with subjective experience to be considered as important as that of neuroscientists. There remains the supposedly simple task of solving how neural patterns correlate with certain states of experience (Solms and Turnbull 2011). In the next section, however, I will explain why this is anything but trivial.

Methods and Scientific Theoretical Position
Psychoanalysis seeks to understand and describe the subjective and individual. It is classified by many authors as a hermeneutic or idiographic science. The neurosciences, on the other hand, are counted among the nomothetic sciences. Their aim is to make universally valid, objectifiable statements about reality. While this rigid classification is not entirely correct – psychoanalysis, for example, also attempts to make universally valid statements about the world – this simplification makes it clear that the two disciplines are in principle committed to very different goals (Warsitz and Küchenhoff 2015). This is also reflected in their methods and conceptualization. The subjective and the unconscious are best captured in a depth hermeneutic analysis of unfolding interactions that may be contained in a detailed case description (Lorenzer 2002). Making generally valid statements, on the other hand, presupposes the objectifiability and falsifiability of clearly definable objects of inquiry (Grünbaum 1985).

2.2 Epistemological Considerations

Neuropsychoanalysis has to engage with and bridge both scientific traditions. This exchange can be realized in different ways. In the simplest case, one sub-discipline takes note of the findings of the other discipline, but only allows itself to be influenced if it seems unavoidable, for example, if the other discipline has achieved such a fundamental gain in knowledge that its own position is simply no longer tenable. It becomes difficult when a concept from one discipline is to be examined with the methods of the other in order to verify its legitimacy, as, for example, in the experimental investigation of defense mechanisms (see Sect. 3.2.2). The conceptual definitions of psychoanalysis are often so vague or multi-layered that a concrete operationalisation in order to test the concept in a scientific experiment runs the risk of rendering the concept devoid of its essence. Thus, the results may be virtually worthless for the study of the actual concept. Concepts of one discipline may be ultimately untranslatable, resulting in an *incommensurability* that has to be bridged in order to enable an exchange between disciplines.

Neuropsychoanalytical Research 3

3.1 Operating Principles of the Mindbrain

In his metapsychological reflections, Freud sought to understand and describe the basic determinants of human experience and behavior (Freud 1900, 1915a, b, c, 1923, 1950 [1895]). He therefore developed concepts about drives as basic sources of energy that control the organism to regulate biological functions, ensure survival, and reproduce. The dynamics of these drives are regulated according to economic principles: An increase in instinctual tension, if this drive cannot be satisfied, leads to displeasure. On the other hand, the pleasure principle is characterized by the quickest and most effective removal of tension resulting from unsatisfied drives. This direct discharge of drive energy is, however, inhibited in humans by external and internalized factors, and must submit to the so-called reality principle. Drive energy must be at least temporarily inhibited, bound, and transformed into a form that enables socially accepted behavior or cognition. This operation signifies the transition from primary to secondary process. This transformation is accompanied by the associated contents becoming conscious or conscious-in-principle (preconscious) and thus, according to Freud's first model of the mind, passing from the Unconscious (or system Ucs.) into the system Pcs.-Cs. This so-called "topographical model" overlays the later developed "structural model" consisting of Id (as the unconscious seat of drives), Ego (as the mediating instance) and Superego (as the seat of norms, ideals and internal objects shaped by social experiences). On the basis of neuroscientific findings on affects and motivational systems in mammals (Panksepp 1998, 2011) and statistical learning models (Carhart-Harris and Friston 2010), Mark Solms developed a fundamental revision of Freudian metapsychology that draws on current findings in neuroscience (Solms 2020).

© The Author(s), under exclusive license to Springer Fachmedien Wiesbaden GmbH, part of Springer Nature 2023
G. T. Waldhauser, *Neuropsychoanalysis*, Springer essentials,
https://doi.org/10.1007/978-3-658-40891-6_3

3.1.1 Drive and Emotion

Since Freud, the biological foundation of drives has been repeatedly discussed and criticized from different perspectives (e.g. Storck 2020). One possibility consists in formulating the drives in purely psychological terms and decoupling them from biology, the other in underpinning Freud's considerations with findings from behavioral biology or revising them on the basis of these. Comparative behavioral research shows that a limited number of emotional and motivational systems have developed in the course of phylogeny, which can be found across various cultures and animal species. Based on animal studies, behavioral biologist Jaak Panksepp differentiated seven distinct motivational systems that fundamentally determine mammalian behavior. The neural correlates of these systems are located in subcortical layers of the brain and are each controlled by specific hormone and neurotransmitter circuits (Panksepp 1998, 2011). These seven systems are named according to their main function and emotional quality:

- The SEEKING system: curiosity, interest, the need to explore and interact with the environment. Motivated seeking behavior is fundamentally important to satisfy basic biological needs such as hunger or thirst, but it is also engaged, for example, to control reproductive behavior. Thus, the SEEKING system has a special position, as it serves the other systems by providing energy. It corresponds to the mesolimbic dopaminergic reward system, which also plays a central role in addiction.
- The LUST system serves the search for sexual partners and reproduction. It is regulated by sexual and bonding hormones.
- The FEAR system serves to protect against physical harm through escape. It influences cognition and behavior through rapid subcortical circuitry across the lateral and central areas of the amygdala and through the release of stress hormones.
- The RAGE system serves to eliminate possible sources of threat or obstacles to satisfying one's own needs and is linked to the amygdala in its medial part and the efferent stria terminalis.
- The PANIC system: Controls behaviors related to the need to form attachments. The separation or loss of attachment figures is experienced as panic and despair.
- The CARE system embraces compassion, concern for others, and is the source of parental care.

3.1 Operating Principles of the Mindbrain

- Finally, the PLAY system corresponds to the need to play, against the background of determining one's own position in the social group, e.g., in playful fights determining dominance hierarchy. The last three systems are essentially controlled by bonding hormones such as oxytocin and opioids.

The individual subsystems are also interconnected in an inhibitory or excitatory manner. For example, confrontation with an overwhelming threat leads to activation of FEAR, which simultaneously inhibits the RAGE system. The motivational systems are neuronally hardwired to a greater or lesser extent and initiate goal-directed behavior. However, this does not occur in an automated and unconscious manner, but is usually associated with consciously experienced affective qualities. This basic awareness of affect arises via the interconnection of all the motivational systems described here with the periaqueductal gray. According to Panksepp, this cell complex forms a subcortical control center that processes and links information from all motivational systems. Building on this theory, Solms argues that by closely linking with the networks of the ascending reticular system, which is fundamental to the regulation of levels of consciousness, a core awareness of the state of the self in the world could emerge here (Solms 2019, 2020). Panksepp (2011) therefore refers to this region as SELF, or core affective self. Through connections with cortical regions, such as the anterior cingulate cortex, these basal motivational circuits gain access to cognitive processes and executive control functions.

Far-reaching conclusions can be drawn from these findings and their interpretation: Based on the data, it is difficult to understand drives in purely psychological terms and to ignore the fact that, beyond the concepts of "libido" and "death drive", there are a number of fundamental drive systems that significantly control experience and behavior. Furthermore, they suggest that consciousness is something fundamental that emerges not only via complex neural circuitry in the neocortex, but already in phylogenetically ancient layers of the brain (Solms 2019, 2020). If one follows this conclusion, however, it also means that an essential statement of Freud's structural model has to be revised. Whereas Freud held that the Id, the seat of drives and archaic behavioral imperatives, was unconscious, Solms takes the opposite view. Rudimentary affect consciousness is a necessary condition for speaking of drives, since it is only through the sensation of the qualities of affect (pleasure, anger, fear, etc.) that the regulatory demands from the organism become behavior-controlling drives. The disappearance of the consciously experienced affects marks the satisfaction of the needs associated with them (Solms 2019, 2020).[1]

3.1.2 Learning Through Experience

In order to achieve satisfaction of needs, the organism interacts with its environment in a way that allows it to minimize feelings of displeasure. As we know from Freud, however, this may be anything but easy and requires a constant confrontation with internal and external demands. Avoidance of displeasure can be optimized by learning to predict internal and external events as well as possible.

The so-called "predictive coding framework" interprets all cognitive processes very comprehensively as a statistical reduction of prediction error. According to this idea, the brain continuously tries to generate the best possible model of the next events. Consequently, prediction error leads to an ongoing adaptation of our heuristics about ourselves and the world to changing environmental conditions. If the prediction error cannot be reduced by model adjustments, we can attempt to interact with the environment in such a way that it changes in line with our expectations. For Solms and others, this notion is quite consistent with Freud's ideas on the discharge of instinctual energy. Predictive error is conceptualized as greater entropy or "free energy" that signals a violation of homeostatic equilibrium and should therefore be minimized (Carhart-Harris and Friston 2010; Fotopoulou and Tsakiris 2017; Solms 2019, 2020).

This somewhat abstract model can easily be illustrated by everyday situations. In the course of our lives, we manage to extract regularities from many individual situations, which form schemas, scripts or world knowledge. These schemas guide our behavior and thinking in situations that are known to us, at least in principle (Moscovitch et al. 2016). We usually do not have to think consciously when we turn on the coffee machine in the morning, because we know our kitchen and the

---

[1] This view is not without problems, however, as it treats the question of affect and consciousness in a simplistic way. It is well known that emotions have both conscious and unconscious components and that the ability to perceive affect is quite significantly related to mental health (Damasio 2003). Solms and Panksepp remain somewhat unclear here as to whether the quality of consciousness from the core self maps the entire spectrum of affect, or whether core consciousness refers to a rather undifferentiated awareness of the basic level of arousal in the organism. Just like the awareness of certain objects in the outside, a more complex and differentiated perception of affects can possibly only arise through interaction with ego structures.

necessary actions from a multitude of similar episodes. Maybe it smells like fresh coffee shortly after, but even that may not be something we notice on an ordinary morning. Only when something in the regular sequence changes, and depending on the size of the prediction error and the significance of the deviation of the events for the organism, do we become aware of sensory impressions. For example, if some coffee powder is spilled on the floor next to the machine, we may not notice. However, if the coffee powder container is unexpectedly empty, or if it suddenly smells like burnt cables instead of coffee, the schematic process is disturbed to such an extent that we become aware of the situation. In the latter case, action is required because there is concrete danger to life and limb. Solutions have to be found, which necessitates thought processes and possibly an inhibition of the immediate impulse to act.

Translated into the neuropsychoanalytic reinterpretation of psychodynamic processes, the olfactory stimulus of the scorched cable signifies a prediction error that triggers the FEAR system, which manifests itself in the conscious affect of startling. The primary action impulse, flinching back, must be suppressed and affective energy has to be linked to cognitive thought processes in working memory – in psychodynamic terms: the primary process drive impulse must be inhibited and transferred to the secondary process (Solms 2019, 2020). By linking the conscious affect to the usually unconscious thought processes, these thought processes become conscious. This is consistent with the idea that cognition only occurs consciously to the extent that it is important to us (Solms 2019, 2020). Secondary-process cognitive binding of the primary-process affective impulse allows a goal-directed action to be planned and executed, e.g., that the developing fire can be averted with an appropriate extinguishing agent, the power turned off, and the cause of the problem identified. This immediate consciously guided behavior initially reduces entropy again, but in order to permanently reduce the prediction error, the internalized models must be "updated". This is done by linking stored long-term memory content with current working memory content and their consolidation or reconsolidation. Therefore, the affect of fear may still be felt for a while when making coffee and one will now consciously check the cable of the machine regularly, but after a while this will also become part of the schema. Far more problematic are situations where entropy cannot be fully resolved. This can be the case when the affect energy cannot be sufficiently resolved through action or the intermediary thought process cannot find solution patterns. This can lead to the affect initially remaining unresolved and thus, conscious, overshadowing what were previously innocuous kitchen situations, and making one uncomfortable at the sight of a coffee machine. In this trivial example, the importance of this model as an explanation for mental disorders becomes clear.

It becomes even more tangible if we consider this in the context of child development, a stage of life in which ego functions are not yet fully developed. Here, it is often possible to temporarily find and consolidate sufficiently good possible solutions to a conflict, but these do not resolve the entire prediction error. As an example, being excluded from the adult world as a child can cause strong feelings of anger, as one feels inferior and set back. Instead of binding this anger in order to use it for one's own development after initial frustration and to arrive in the adult world in the long run, it may be temporarily more obvious to show oneself quite childlike and needy. By this, a way is found to be with the adults after all, since they have to take extra care. This approach serves as a template for later situations of social exclusion and is stored away as a viable heuristic. However, this approach leaves some free energy unbound – the original feelings of inferiority and anger are not permanently alleviated, but rather possibly intensified. The original anger may now be directed against the self, for example, when a person somatizes and becomes physically ill in similar situations in the future. Alternatively, the anger may remain unbound and reappear in the future in similar moments of social threat in raw form, for example in states of panic (Solms 2018).

Solms conflates premature storage of suboptimal solution attempts with Freud's primal repression (Solms 2020). However, before turning to the neuroscientific findings on repression and defense mechanisms, it is useful to take a closer look at the neuroanatomical correlates of basic mental functions and possible cross-connections between the structural model of the psyche and different brain areas.

3.1.3 The Neural Basis of the Structural Model

Which brain areas are responsible for processing affects and sensory signals in such a way that goal-directed, drive-economic actions emerge from them? As outlined above, the intrapsychic stimulus and energy sources, the affect systems, the homeostatic regulatory circuits, and the control of the basic level of wakefulness and consciousness can be located in subcortical and limbic structures. These structures project to the medial parts of the cerebrum, the cingulate gyrus and the medial temporal lobe, summarized as the limbic cortex. These projection routes thus represent a perceptual system directed toward the interior of the organism. Sensory stimuli coming from outside, on the other hand, are processed in the unimodal sensory areas in the occipital (vision) and temporal (hearing) cortex, after stimulus transmission has already been preprocessed and influenced by deep brain structures such as the thalamus and the amygdala. According to Solms, these structures represent the *Pcpt.-Cs.* system postulated by Freud, and include the ω-neurons

3.1 Operating Principles of the Mindbrain

(Freud 1900, 1950 [1895]). The basic perceptual information is now gradually linked together. Along the ventral and dorsal visual pathways, the identity of perceived objects and their localization in space, respectively, are represented until this information is finally merged in the multimodal association areas in the parietal lobe.

Right-sided lesions in this area, especially in the temporoparietal junction, are often accompanied by a hemineglect, a disturbance of attention and control for the environment and side of the body opposite to the damaged brain area. Kaplan-Solms & Solms add to the description of purely cognitive deficits and report from the psychoanalytic treatment of neglect patients (Kaplan-Solms and Solms 2000). Some of the cases presented are unaware of their deficits in terms of a lack of insight into the illness (anosognosia). They can, however, become temporarily aware of their deficits after appropriate interventions, but they find this emotionally extremely distressing. Other patients, on the other hand, perceive the half of the body that can no longer be controlled as seemingly no longer belonging to them and would prefer to destroy these limbs. They identify this half of the body, as well as the surrounding environment, as the cause of their own misery. The authors interpret these emotional consequences of the neglect as the result of defense mechanisms that serve to maintain the self-love that has arisen from a libidinous cathexis of one's own body. If this body as a self-object can no longer be experienced and controlled as a whole as a result of the brain damage, the bad and injured parts must be split off, repressed or projected outwards.

A crucial step in the further processing of external and internal stimuli is to link them with other, current stimuli or memory representations, embed them in a semantic network and allow for symbolization. These processing steps are closely linked to language in the human brain. Patients with left hemisphere damage have deficits not only with comprehension and production of language, but also, as Solms and Kaplan-Solms suggest in line with working memory studies (Baddeley 2003), with the transfer of thought into consciousness. What is essential in this step is that not only the representational format changes, but also the processing mode. Kaplan-Solms and Solms assume that perception and the processing of different modalities and stimulus qualities take place in sensory and associative areas in a massively parallel fashion, whereas language and the goal-oriented sequencing of working memory content and actions require serial processing. In Freudian terminology, this latter type of processing corresponds to the *secondary process* (Kaplan-Solms and Solms 2000).

Sequences, prioritizations, and schemas that govern actions and cognitive processes are shaped by learning experiences and are physically represented in the prefrontal cortex (PFC). Taken together, the PFC corresponds to the system *Pcs.*-

Cs., it is the seat of higher ego functions, thought processes, attentional control, and central defense mechanisms. An important function of the PFC is the binding of cognitive content and its inhibition in order to enable behavior that reduces the prediction error as effectively as possible and thus meets both inner needs and external requirements.

The PFC is the part of the brain that takes longest to develop over the lifespan. It is hierarchically structured along a caudal-to-rostral axis in the complexity of its functioning (Badre and Nee 2018). At the same time, the PFC can also be described in dorsal-to-ventral and lateral-to-medial gradients. While cognitive control mechanisms are localized in the dorsolateral regions, the ventromedial area is closely related to the appraisal, inhibition, and control of emotions. It becomes active when we think about ourselves, but also when we think about emotionally close people (de Waal and Preston 2017). In addition, it is also the consolidation endpoint of the episodic memory system, i.e., abstract regularities between individual autobiographical events are stored here in an extracted manner and reactivated as needed (Moscovitch et al. 2016). At the same time, this brain region has direct reciprocal connections to the subcortical structures of the emotion systems. The ventromedial PFC exerts an inhibitory influence on subcortical regions and significantly influences the network of prefrontal control regions (Kringelbach and Rolls 2004).

Kaplan-Solms and Solms (2000) report from the psychoanalytic treatment of several patients with different damages in this region. Even if the cases differ drastically in their superficial symptoms – some show uncontrolled, impulsive and disinhibited behavior, others a flattening of affect – a psychoanalytic point of view reveals fundamental similarities. The way of thinking and feeling in these patients shows the characteristics of the system *Ucs*. Thus, these patients do not know contradictions. Patient H., presented by the authors, is aware that his son has long since died, while at the same moment reporting his recent visit. This absence of contradictions is coupled with marked wishful thinking. Mr. H. imagines himself not in a neurological clinic but on a cruise ship in the Caribbean. This happens especially when the patient momentarily realizes his current situation. The reality principle is replaced by the pleasure principle and external reality is replaced by internal reality; hallucinations and psychotic ideation are also reported. The patients also show a certain timelessness of thought, corresponding with Freud's ideas on the *Ucs*. Mr. H., for example, could never tell the time, or according to his (wishful) conviction it was always visiting or eating time. The temporal succession of events from his life was also confused. Finally, the patient's thinking and feeling show clear traits of the primary process: affects and ideas seem disconnected, fluid, and behavior disinhibited. These patterns seem to be accompanied by an absence of

3.2 Memory and Defense

internal objects, which is evident in the transference relationship. The therapist feels that she is acting as an auxiliary ego and as a structuring or holding object. However, due to the lesion, there is no consolidation of these functions beyond the therapy sessions.

In summary, the PFC can be identified as the seat of higher ego functions. Together with medial parietal structures, the right temporoparietal junction described above, the anterior cingulate cortex and deeper brain structures, the medial part of the PFC forms the core of the so-called default mode network, which is central to the experience and regulation of the self (Sect. 3.3). The ventromedial part shows the properties of the superego described by Freud: it is the latest and longest maturing part of the ego, arises via the internalization of early object relations, and simultaneously interacts with the subcortical neuronal substrates of the ego to inhibit the primary process (Kaplan-Solms and Solms 2000; Freud 1923).

The organization and functioning of the brain can thus be described from a psychodynamic perspective. Of course, it can be noted that cognitive approaches can provide far more elaborate concepts and empirical evidence for the functioning of individual brain regions than a theory that is over 100 years old. However, neuroscientific theories of individual cognitive and affective domains are often isolated and lack a synopsis of individual basic functions. Psychoanalytic practice and thinking, however, requires precisely this synopsis and thus, attempts to find overarching functional principles for the brain can benefit from a depth psychological approach.

3.2 Memory and Defense

As explained above, the psychoanalytic approach converges with theoretical concepts of statistical learning models. The predictive coding framework is in principle applicable to any level of observation, from single neurons, to the whole brain, to the psychoanalytic couple. A key feature of each subsystem is its ability to predict events in order to reduce statistical error and thus entropy. A precondition for this, in turn, is the ability of the system to process information not only at a particular moment, but also its permanent plasticity, i.e., the property of being changed by previous stimulation or processing. Only then can an update of the internal models occur and be used in the future. Equally important in this context is the ability to temporarily inhibit model predictions taking into account current demands that have to be processed in the system. If this inhibition fails, thinking and behavior patterns occur reminiscent of patients with ventromedial damage. These processes of storage, reactivation and inhibition will be described in the following section on memory and defense.

3.2.1 Memory Systems and Memory Processes

Human memory can be divided into different subsystems with different properties. This has important consequences for the interpretability of different memory phenomena and also for the perspective of being able to change the system through biological or psychological interventions.

Memory Systems
Basically, human memory can be divided according to the degree of awareness with which information is retrieved (Squire 2004; Lane et al. 2015). For *implicit* memory, remembering takes place unconsciously; for example, certain actions or perceptual processes can be performed more quickly if they have already been carried out (procedural memory) or certain objects have already been seen (priming). Implicit memory also involves associative learning of emotional links (conditioning). For example, you may sweat and tense your muscles in particular situations, that have been associated with a fear-eliciting experience, even if you don't remember being in a similar situation.

Explicit memory can be divided into *episodic* and *semantic* memory. Semantic memory includes knowledge about the world and facts about one's self and life. One consciously remembers the name of the city of one's birth or can describe one's kitchen. Episodic memory involves remembering specific events from the personal past. Episodic memory involves an "autonoetic" awareness of one's self in time and space: one remembers the place and time of the original experience with the help of a "mental time travel" and can separate the past from the present in the process.

The different systems are also associated with different neural structures. Episodic memory depends largely on the medial temporal lobe, including the hippocampus, which interacts with control areas in the lateral prefrontal and parietal cortex during storage and retrieval. Essential structures for semantic memory are found in the temporal lobe, particularly at the anterior temporal pole, and in the left inferior PFC in partial overlap with important language areas. The core structures of implicit memory are highly heterogeneous and dependent on the specific domain. While priming of perceptual patterns occurs in the neocortex, classical conditioning occurs in deeper brain structures such as the amygdala.

Memory Processes

Despite the separation of the different memory systems described above, the different systems interact with each other. The affective tone of an event has a significant influence on how a memory is stored in episodic memory. The amygdala influences the functioning of the hippocampus via the release of stress hormones (Cahill and McGaugh 1998). Moderate emotional stimulation leads to improvement and extreme stress leads to fragmentation during consolidation. Consolidation refers to the process of slowly solidifying memory content. The process of consolidation also leads to a transformation of memory content (Moscovitch et al. 2016). Thus, episodic memories are decontextualized and pass into semantic memory when similar episodes are frequently experienced. Regularities between different episodes are extracted and crystallize into semantic schemas. The different memory systems also interact with each other when remembering autobiographical events (Lane et al. 2015). For example, our schematic knowledge of what a kitchen usually looks like controls our very personal memory of what happened this morning when the coffee machine went up in flames. As we remember, the affect of fear is also reactivated, until we realize that the reason we remember the event is because it smells of something harmless on the street.

The consolidation of autobiographical memories is not a singular event. Each retrieval of a memory labilizes the existing memory traces and leads to a renewed consolidation ("reconsolidation"; Lane et al. 2015; Nader and Hardt 2009). On the one hand, this can lead to a consolidation of already established traces; on the other hand, each retrieval situation also functions as a new episode and leads to the storage of new memory entries that are associated with the original episode and can thus lead to its permanent change (Lane et al. 2015; Nader and Hardt 2009).

Reconsolidation as a Mechanism of Action of Psychotherapy

Reconsolidation provides a model for the persistence of mental disorders and at the same time for the mechanisms of action of psychotherapy. On the one hand, experiences that are made or remembered over and over again solidify and schematize. On the other hand, the notion of reconsolidation suggests that the templates through which we perceive the world are in principle accessible and modifiable. If this modification takes place, be it through targeted conscious recollection of autobiographical episodes, questioning cognitive beliefs or perceiving and working through emotions in a protected setting, existing memories can be enriched and corrected with new learning experiences. Here, high-frequency, long-term psychoanalytic therapy creates a unique framework for accessing even deeply rooted memory traces (Lane et al. 2015). The continuous linking of mental content in the analytic process creates an effective and identifiable network of memory stimuli and representations. The high frequency and setting of psychoanalytic therapy facilitates mental state changes, potentially enabling the recollection (or reconstruction) of remote autobiographical memories (Koukkou and Lehmann 1998; Mendelsohn et al. 2008). Working with transference allows existing schemas and implicit memories to be experienced and reinterpreted with the help of the therapist, who is emotionally attentive and trained through many years of self-experience (Lane et al. 2015). The resulting new and corrective experiences can be stored along with the original memory traces and modify existing memories, schemas, and implicit links. Much research would be needed to confirm these possible effects of psychoanalysis on memory. However, it must be acknowledged that the mode of action of psychoanalysis is in principle consistent with theoretical models of basic memory processes.

3.2.2 Repression and Defense Mechanisms

For all its agreement in principle with modern theories of memory, however, psychoanalysis also postulates the existence of repression, a memory mechanism that is still subject of highly controversial debate today. Freud assumed that thoughts associated with unsolvable psychological conflicts are banished from consciousness and thus can no longer be consciously remembered (Freud 1915b). This occurs in two steps: During the stage of so-called *primal repression*, the individual comes into this conflict for the first time. The cognitive contents (perceptual patterns, thoughts) associated with the conflict are deprived of their energetic cathexis, thus becoming unconscious. However, the affect related to this conflict remains. Freud assumed that the repressed contents continue to persist in the Unconscious and strive to re-enter consciousness. Continued repression can only occur through continuous inhibitory effort and through the counter-cathexis of sufficiently similar, but non-conflict related, "screen memories". When the individual is confronted with thoughts and events that resemble the original repressed conflict material, the repressed material strives to become conscious again. As a result, the new material must also be repressed. This second step is referred to by Freud as *after-expulsion*. In the third step, there is a return of the repressed in the form of behaviors, symptoms, and dreams through which the repressed seeks to gain access to consciousness. The potentially pathogenic effect of the repressed highlights the urgency of recalling repressed material and working through it in a therapeutic context (Freud 1914a; Lane et al. 2015; see Grünbaum 1985).

3.2 Memory and Defense

Primal Repression and False Memories

It is almost impossible to judge whether a memory is veridical or not. It is extremely easy to falsify and construct memory content, especially in interaction with other people (Loftus 2017). This is especially true when memory content is stored in a format that limits conscious access. In the 1980s, the debate about the veracity of repressed memories culminated in the so-called 'memory wars'. Cases of alleged child abuse, for which memories had faded for many years and which then suddenly were remembered again, often in the context of suggestive therapy procedures, were tried in court. Through external evidence and experiments demonstrating the fragility of memories, it became clear that many of these cases did not correspond to reality (Loftus 2017).

Freud assumed that the inability of humans to remember events from early childhood was due to the barrier of primal repression (Freud 1915b). This implies that events become rememberable when this barrier is dissolved. However, this theory can be considered outdated and false. The structures essential for episodic memory, the hippocampus and parts of the PFC, do not mature until the age of three and continue to function differently from adult brains well into school age (Alberini and Travaglia 2017). All episodic memories from early childhood are therefore, in all likelihood, largely constructions. These constructions are also, to a significant extent, enriched with details that correspond to social desirability and narratives that emerge, for example, in therapy. Remembering, consolidation, and reconsolidation are intersubjectively constructed, transformative processes.

However, unconscious memory traces, which can never become conscious, can also exert an influence on experience and behavior in the form of schemas or implicit behavioral dispositions, which are recognised and made tangible in psychoanalytic therapy (Lorenzer 2002). This, however, is not an active case of repression. Solms (2018, 2020) assumes that primal repression describes the process in which schemas and behavioral scripts are prematurely stored when they seem to ensure the best possible conflict resolution. However, no continuous effort is required to prevent these memories from becoming conscious. Their unconscious persistence may rather arise from the fact that the originally stored schema has proven to be sustainable in dealing with psychological challenges over a long period of time and is therefore robust against attempts to change it. According to Solms, the "return of the repressed" refers not to the "repressed" content itself, but to the affects associated with the conflict. Due to the limited power of the heuristic and the thus not fully reduced prediction error, a portion of the affect energy remains unbound and thus may become conscious again and again.

Defense Mechanisms and Inhibitory Control

But what about when an individual experiences a psychological conflict, cannot fully resolve it, and already has a sufficiently well-developed brain? How can he or she deal with this experience? Neuroscientific memory research shows that encoding and remembering are dynamically controlled processes that can be influenced by motivational aspects and controlled consciously or unconsciously.

Already in the early days of psychoanalysis there were attempts to investigate repression experimentally. Carl Jung presented his subjects with neutral and conflict- or symptom-related words, whereupon they were asked to associate freely (Jung and Ricklin 1906). During this association phase, skin conductance and reaction time to the onset of the first association were measured. At a later time, the subjects were presented again with the stimulus words and they were then asked to reproduce their original associations. Jung and colleagues found that associations to symptom- or conflict-related words were generated after a longer reaction time, resulted in increased skin conductance (indicating an increased physiological stress response), and were more likely to be forgotten than emotionally neutral words (Levinger and Clark 1961). A major problem with these studies was the supposedly inadequately controlled stimulus material. The same criticism has been brought forward against studies of so-called "perceptual defense". These experiments, developed from projective tests and sublimimal perception research, showed that conflict-related material was perceived later than emotionally neutral material (Dixon and Lear 1963; Kragh 1960). In all these experiments, however, it was also unclear which cognitive or emotional subprocesses might be responsible for a possible defensive process.

Thus, attempts were soon made to define and operationalize repression and defense more narrowly. Initially, "displeasure" caused by psychological conflict was reinterpreted as "emotionally negative" and recall of well-controlled, emotionally negative stimulus material was tested. However, it was found that recall of emotionally negative material was generally easier, regardless of emotional valence (Rapaport 1971; cf. Cahill and McGaugh 1998). Studies employing the induction of shame or electric shocks in order to establish "ego-threat" to prevent the recall of certain memory representations again were unable to rule out the confounding influence of other variables (e.g., social desirability or fear of painful stimulation during recall; Weiner 1968; Zeller 1950).

In order to better delimit and control the mechanisms of motivated forgetting, an essential intermediate step was introduced. In the tradition of so-called "goal-directed forgetting", forgetting was not operationalized as an automatic consequence of possibly stressful experiences, but via the intermediate step of a conscious decision to forget certain events and thoughts. In these experiments, after the

3.2 Memory and Defense

presentation of individual stimuli or an entire list of stimuli, subjects are instructed to forget the material just shown (Anderson and Hanslmayr 2014). The "directed-forgetting" effect measured in this way is robust: to-be-forgotten material is actually remembered worse than to-be-remembered material. Some experiments have also been able to show that the forgotten material is not deleted, but can be recalled by restoring access routes, which can be interpreted as a removal of retrieval inhibition (Geiselman et al. 1983). Recent neuroscience studies emphasize the role of neural inhibition in goal-directed forgetting. Instructions to forget lead to the activation of cognitive control functions in the PFC and to an amplitude increase of inhibitory, so-called alpha oscillations over brain regions that process the to-be-forgotten stimuli (Fellner et al. 2020).

The most recent experimental development in this area is the so-called "think/no-think" paradigm (Anderson and Green 2001). Here, a direct instruction to forget a particular stimulus is no longer given during learning. After an initial phase of studying of pairs of words or pictures, one member of the pair is presented repeatedly in a second phase. In the so-called think condition, subjects are asked to practice recall of the associated other pair member. In the no-think condition, the other member of the pair is to be deliberately suppressed. Suppressed no-think material is recalled worse in a surprise memory test than material that was shown during study but neither recalled nor suppressed during the second phase. Again, neuroimaging and EEG studies show increased activity in cognitive control regions such as the dorsolateral PFC and anterior cingulate cortex and inhibition of memory areas such as the hippocampus (Anderson et al. 2004) and a reduction in theta and gamma oscillations associated with sensory memory retrieval (Waldhauser et al. 2015, 2018). The general ability to suppress stressful material appears to have a protective effect against the development of post-traumatic stress disorder (Mary et al. 2020; Waldhauser et al. 2018).

The question is whether the empirical evidence for conscious supression and directed forgetting has any implications for the empirical plausibility of repression. Several arguments have been brought forward that conscious suppression is not fundamentally separate from unconscious repression. First, observing an unconscious mechanism in the here and now says nothing about whether the mechanism was also unconscious or conscious at the time it arose. Other skills and cognitive functions are also initially performed consciously and then become increasingly automated, so there may not be a truly sharp boundary between suppression and repression (Brenner 1973; Dollard and Miller 1950; Erdelyi 1985, 2006). Second, empirical findings show that suppression in the think/no-think paradigm can occur unconsciously when triggered by subliminal stimulation (Salvador et al. 2018). Third, the underlying mechanisms of suppression are highly similar to unconscious forms of controlled forgetting, particularly retrieval-induced forgetting. This line of research examines how information can be selectively retrieved from memory against the interference of similar memories. As robustly shown in a multitude of studies, interfering memories are dampened and forgotten in the process. Again, neuroimaging and EEG studies reveal the neural signature of inhibitory processes, higher activity in prefrontal control regions, and inhibition of areas that house interfering memory traces (Waldhauser et al. 2012). Thus, there appears to be an underlying network for the top-down control of memory content that can be controlled both consciously and unconsciously.

Against this background, recent studies have made further attempts to use stimulus material that is as naturalistic as possible to investigate the neural correlates of defense mechanisms. In a study by Shevrin et al. (2013), subjects were subliminally presented with key words from their psychoanalytic therapy sessions. Compared to keywords from other subjects, there was an increase in alpha oscillations associated with prolonged reaction times and higher physiological arousal. The interpretation of this brain oscillatory pattern as evidence for unconscious inhibition is consistent with both recent and several decades old EEG studies (Bazan 2017; Dixon and Lear 1963; Fellner et al. 2020; Waldhauser et al. 2012). Schmeing et al. (2013) ported Jung's association experiment to the MRI scanner. Instead of the original word material, they adapted prototypical sentences representing the most frequent psychodynamic conflicts according to the Operationalized Psychodynamic Diagnosis system (Arbeitskreis OPD 2006), which were also controlled for linguistic and affective criteria. Compared to neutral and emotionally negative material, conflict-related stimuli elicit longer reaction times, more forgetting, increased skin conductance, and activation of cognitive control areas that also become active in the think/no-think paradigm and retrieval-induced forgetting. At the same time, there was a reduction of activity in the hippocampus, which indicates an effective inhibition mechanism (Schmeing et al. 2013).

In summary, there are neurocognitive mechanisms for inhibiting retrieval and storage in memory that are active when memory content is maladaptive to achieving momentary goals, whether it is remembering other information or making a voluntary decision not to think about something. Operationalizations using decidedly psychodynamic material appear to address similar mechanisms as the far more numerous and better studied approaches to directed and retrieval-induced forgetting.

3.3 Developmental and Social Psychological Perspectives

The previous sections have touched on which structures and processes in the "mindbrain" serve the organism to regulate itself in constant interaction with the environment. One of the most central insights of psychoanalysis is that this system undergoes a fragile development that is significantly shaped by early experiences with other individuals. Whereas Freud's focus was still on the intrapsychic determinants of development and their later conflict with the demands of external reality, object relations theory has questioned the assumption that human beings can actually be thought of as individual beings at all in their early development. The human organism is not fully developed at birth and is thus dependent on others for its survival. The abilities to self-regulate and to understand oneself and others emotionally and cognitively are developed and internalized through close interaction with caregivers (Bion 1962; Kohut 1983; Winnicott 1960). To what extent do these assumptions coincide with the findings of neuroscience?

The brain is designed from birth to learn and process social stimuli as effectively as possible. Thus, an excess of synaptic connections develops in the first two years of life. The resulting "overconnectivity" enables an initial rapid adaptation to all conceivable environments and learning experiences (Sakai 2020). Over the course of development, redundant neurons are reduced in phases, with a particularly strong elimination phase at the onset of adolescence. Through this mechanism, networks are formed that are enormously effective and robustly adapted to previously experienced environmental conditions. Subcortical structures are already largely developed at birth, but the cerebral cortex is only rudimentarily defined. However, sensory brain areas are already developed to the extent that they allow preferential processing of social stimuli – faces, voices and human speech sounds. These areas form in the first months of life so that specific faces or voices are discriminated from others and are processed preferentially (Parsons et al. 2010).

A crucial developmental step is the functional formation of the default mode network (Sect. 3.1.3). This network serves to link internal and external information through the interaction of different subcortical and cortical areas. It becomes active when we think about ourselves, daydream, remember our personal past, and when we plan our future. A substantial part of the default mode network is located in the medial PFC, which we have already met in the discussion of the localization of the Freudian structural model as the seat of the ego and superego functions. As this region matures, the child begins to exhibit goal-directed behavior and increasingly develops a "theory of mind," becoming aware of both his or her own mental states and the states of others (Frith and Frith 2012; de Waal and Preston 2017). Other major nodes of the default mode network are located in medial posterior areas such as the precuneus and posterior cingulate cortex. These areas internally integrate information from deeper brain regions and interact with memory regions in the medial temporal lobe and hippocampal formation. Related to the neural maturation of this network is the formation of a coherent self-experience and autobiographical memory system (Alberini and Travaglia 2017; Moscovitch et al. 2016). Also important in the context of the development of self-experience is the development of the temporoparietal junction. As already seen in neurological patients, this region is important for experiencing a coherent self (Fotopoulou and Tsakiris 2017). As this brain region matures, the child is able to recognize himself in the mirror (Parsons et al. 2010; Lacan 1948). Like the medial prefrontal structure, this brain region also has a dual role: not only does it serve to recognize the self, but it is equally essential for detecting the feelings and mental states of others, making it one of the essential "mirror regions" in the human brain (Frith and Frith 2012; de Waal and Preston 2017).

The maturation of the default mode network begins to unfold in the second year of life and corresponds with associated affective, metacognitive, and self-regulatory abilities (Parsons et al. 2010). However, the medial PFC continues to mature into the third decade of life and provides the neural basis for the development of further metacognitive and social skills. As outlined above, damage to this area leads to massive deficits in regulation of emotion and affect experience. Early neglect and trauma have tangible effects on the formation of these brain regions. Physical abuse reduces the volume of the ventromedial PFC in concert with structural and neuroregulatory changes in subcortical regions such as the amygdala and along the HPA axis (Davidson and McEwen 2012). These changes correlate with a deterioration in social and cognitive abilities across the lifespan.

Far more difficult to answer is the question of how a sufficiently good development of these brain regions can be fostered. Entirely in line with psychoanalytic thinkers (Bion 1962; Kohut 1983; Winnicott 1960), it also seems conclusive from a neuroscientific point of view that the foundation for such development is laid by the early, intuitive exchange between caregiver and child (Parsons et al. 2010). The interaction between caregivers and infants is subject to a finely tuned regulatory circuit that is influenced by the individual brain systems of both infant and caregiver (Rilling and Young 2014). In the mother's brain, hearing infant crying influences the mesolimbic dopamine system (the SEEKING system) and the empathic perception of the infant's state in the insular cortex, which also plays an essential role in the perception of one's own interoceptive signals. Depressed or substance-dependent parents show less mesolimbic or insular responsivity to crying. Overactivity in these systems leads to intrusive parental behavior and stress, which in turn affects infant

3.3 Developmental and Social Psychological Perspectives

behavior and experience. Therefore, a major factor in this interplay are the self-regulatory abilities of parents, represented in the ventromedial PFC, which are significantly shaped by their own early childhood experiences. Oxytocin, which is correlated with a secure attachment style, plays a central role in the modulation of these systems. High levels of oxytocin lead to higher dopamine release (more SEEKING), less stress-induced amygdala activation (less RAGE and FEAR), and higher empathy (CARE) and more effective self-regulatory strategies when confronted with stimuli related to infant stress. This lays the foundation for an intuitive interaction between parent and infant, which influences oxytocin balance and in turn infant attachment style (Rilling and Young 2014). This intuitive interaction has been described in detail by infant researchers (Papoušek and Papoušek 1987; Stern 2010).

Assumptions about the effects of these early interactions on self-regulatory capacity are one of the cornerstones of modern psychoanalytic theory and treatment technique when talking about holding, containing, and mirroring, through which the infant or patient learns to bear and mentalize his or her own affective and somatic states (Bion 1962; Kohut 1983; Winnicott 1960). Mentalization theories describe this process as one determined by higher cognitive-affective abilities (Fonagy et al. 2007), in which the caregiver reads out the infant's affective states, via the mirror systems in the ventromedial PFC and temporoparietal junction, and empathically processes them in the insular cortex, mentalizes them, and makes them available again to the infant in processed form through the parent's own self-regulatory system. Fotopoulou and Tsakiris (2017) argue that much more basal, bodily patterns underly this complex regulative interaction that make the higher regulative mentalization processes possible in the first place. According to the authors, self-experience is shaped in fundamental ways by bodily, synchronized interactions with close caregivers. Fotopoulou and Tsakiris argue that interoceptive stimulus patterns, such as pain or hunger, can be physiologically interconnected with and modulated by external stimulus patterns, such as maternal touch, via specialized nerve fiber connections. Thus, maternal touch reduces the physiological stress response to internal stimuli. Through temporal contingency and direct physiological connections, these two stimulus patterns and their synchronization become associated and internalized. At the same time, maternal touch helps to experience the limits of one's own body and is therefore fundamental to the development of one's self-experience (cf., Anzieu 1996). The experience that distress caused by inner stimuli can be eliminated by specific actions of another person makes it possible to realize that these overwhelming stimulus patterns are in principle changeable. Through frustrating experiences and the development of one's self perception, these functions become increasingly attributable to the self (Fotopoulou and Tsakiris 2017).

Transference and Reactivation of Early Relationship Experiences
Psychoanalysis claims to make early relational experiences at least partially accessible again in the analytic situation via transference processes and thereby to be able to change them. Transference can be understood as a memory phenomenon: In the therapeutic context, implicit and explicit memories and associated states of the self and fantasies from formative social relationships are reactivated through cue stimuli or emotional states. The phenomenon of transference in this conception has been repeatedly demonstrated in social psychological experiments. Perceptions of faces or personality descriptions reminiscent of important people from one's own life lead to attributing characteristics of the reference persons to these previously unknown persons and to feeling as if one were in contact with the respective reference persons (Andersen and Przybylinski 2012). Research on the neural correlates of this phenomenon from neuroimaging studies has remained unpublished so far (Gerber and Peterson 2006).

However, the concept of transference and the mechanisms associated with it are not exhausted by this simple reactivation. Rather, in the psychoanalytic process there is a meshwork of transference of the patient and countertransference of the therapist and phenomena that are difficult to disentangle, such as projective identification, forms of unconscious reenactment and regressive states of consciousness. It can be hypothesized that "transference neurosis" arises essentially via an entanglement of neural mirror systems in the ventromedial PFC, the temporoparietal junction, and, at a very basal and unconscious level, in the motor mirror neurons of the individuals involved (Frith and Frith 2012; Gallese 2009; de Waal and Preston 2017). These processes are unlikely to be mapped "in the scanner" in their full form in the foreseeable future; if anything, they currently remain accessible to the psychoanalytic method alone. However, it would be a step forward if hypotheses could be derived from these findings that could be investigated in the course of therapy, for example, whether self-regulation abilities improve through working through transference and whether this is accompanied by structural and functional changes in the corresponding brain regions.

3.4 Neuroscientific Therapy Research

For a long timee, psychoanalysis has closed itself off from the empirical investigation of its effectiveness beyond case studies. As can be seen in today's debate about health insurance funding and training curricula, this has been very much to the disadvantage of psychoanalysis, even if the arguments against larger-scale effectiveness studies (such as the immeasurability of psychoanalytic processes of change, research as an irresponsible intervention in the therapeutic framework, etc.) are still put forward and certainly need to be considered. Fortunately, however, a rather considerable number of studies have accumulated in recent years that demonstrate the effectiveness of psychodynamic therapy (Leichsenring et al. 2015).

Imaging studies show that psychoanalytic therapies lead to a "normalization" of brain activity in people with anxiety disorders, psychosomatic illness, borderline personality disorder, and depression compared to a healthy control group (Abbass et al. 2014). Neural changes were associated with symptom improvements and a change in test scores in behavioral studies. In a study by Buchheim et al. (2012), a projective interview procedure was conducted with depressed female patients and a non-depressed control group to determine attachment style. Here, the participants saw pictures showing social interactions and they were then asked to express fantasies and ideas about what kind of story was being told in these pictures. A few weeks later, the pictures were shown several times in the fMRI scanner, either together with a personal central sentence from the original interview or with a neutral sentence. There was higher activity in amygdala, anterior cingulate and medial PFC in the personal versus the neutral condition in the depression patients compared to the control participants. After 15 months of psychoanalytic psychotherapy at a frequency of 2–4 times per week, this difference disappeared and the degree of normalization in the anterior cingulate cortex and PFC correlated negatively with symptom severity. These changes suggest improved emotion regulation strategies (Buchheim et al. 2012). Similar results were shown in other studies using individualized stimulus material from the OPD to assess coping with interpersonal conflict (Wiswede et al. 2014).

To date, there are a relatively small number of imaging studies with small sample sizes, varying dependent measures, and heterogeneous treatment procedures (length and frequency of psychodynamic therapies). Nevertheless, the studies convergently indicate that the efficacy of psychodynamic treatments can be demonstrated at the level of behavior and experience as well as at the neural level. In particular, the approach of combining individualized behavioral tests with clear psychodynamic hypotheses regarding conflicts, relationships, and fantasies, with functional imaging procedures seems very promising.

Perspectives and Limitations 4

This text shows that many psychoanalytic ideas are consistent with neuroscientific findings and that some of them could also be verified in scientific experiments. Despite these actually positive findings, there is scepticism, even open rejection, within the psychoanalytic community towards a combination of psychoanalysis and neuroscience. Two central arguments are cited in the following (Blass and Carmeli 2007; Yovell et al. 2015).

First, psychoanalysts do not have to know anything about neuroscience in order to practice psychoanalysis. The mind depends on the biological substrate, but understanding the former is essentially independent of understanding the latter. Of course, one can and must deal with a patient's individual autobiographical memories as they are presented in therapy, for example, even if one has no idea about the neural basis of memory. However, knowing that perhaps certain memory systems were not fully developed at the time of the experience changes the interpretation and therapeutic approach. Yovell describes how, in the treatment of one patient, this insight led to making the patient's unconscious desires and fears accessible in the relationship, rather than aiming at reconstructing supposedly repressed childhood memories (Yovell et al. 2015).

Second, it is argued that neuropsychoanalysis is not genuinely psychoanalytic and that the neuroscientific approach cannot even come close to answering truly psychoanalytic questions about meaning and truth. An orientation toward the neurosciences might lead to a kind of unwarranted or harmful shortcut to psychoanalytic thinking. There is indeed a danger of avoiding the strenuous path of looking more deeply into psychic content and „explaining it away" by changes in neurotransmitter balance or brain structure. However, it does not seem compelling that an engagement with the neural basis of experience necessarily questions the psychoanalytic stance, as implied by this argument. Yovell et al. (2015) argue that the neuropsychoanalytic approach can be a sub-discipline of psychoanalysis without questioning the legitimacy of the hermeneutic method.

Psychoanalysis and First-Person Neuroscience
The neuroscientific investigation of psychoanalytic concepts may involve an unacceptable mutilation of the object of investigation: The subjective experience of an individual. This is due in large part to the methods of measurement, which make it necessary to control experimental and comparison conditions and to repeat stimulations frequently in order to ensure a sufficiently good signal-to-noise ratio in a neuroscience experiment. However, multivariate analysis methods, pattern recognition algorithms, and the development and widespread application of deep neural networks are making it increasingly possible to identify even individual and singular mental representations in the brain.

The enormous potential of this change can be briefly explained using the example of dream research. While 40 years ago it seemed as if neuroscience would reveal that dreams are nothing more than meaningless neuronal discharges (Hobson 2009), it could be shown via experimental research and lesion studies that dreams actually have something to do with libido and less with chaotic electric discharges during REM sleep (Solms 2000) and that the brain is in a regressive state during dreaming that resembles the correlates of primary-process thinking of schizophrenics (Dresler et al. 2015). The neuropsychoanalytic silver bullet would now be to identify the neural signature of individual dream content and perhaps even track it over the course of therapy (Fischmann et al. 2013). Recent approaches show that this is possible in principle. Horikawa et al. (2013) were able to train a machine-learning algorithm on brain patterns recorded while their subjects were looking at everyday objects. This trained algorithm, applied to the brain patterns during sleep, could decode what the subjects had been dreaming. Through these technical advances, the explanatory gap between the mind and brain aspects can be decisively reduced.

Not only can the aforementioned technological innovations in artificial intelligence serve to improve our neuroscientific methods, but the development of neural networks and adaptive machine algorithms may also benefit from a psychoanalytic approach. Algorithms, while inspired by the human brain, often ultimately function atheoretically or via circuitry developed through linguistic or cognitive theories. However, within the machine learning community, the limitations of this approach are being debated and there is growing interest in creating neural networks inspired by basic determinants of human behavior (Hassabis et al. 2017). It is highly speculative, but perhaps it is the technologized twenty-first century that is bringing about a renaissance of psychoanalysis that evolves via neuroscience, and quite differently than Freud and his followers would have dreamed.

What You Learned from This *essential*

- Many psychoanalytic concepts are supported by findings and theories of cognitive, social and affective neuroscience
- The exchange between the neurosciences and psychoanalysis opens up the chance to expand psychoanalytic theorizing and make it compatible with academic psychology again
- The neuropsychoanalytic approach provides important insights for the understanding of experience and behavior

References

Abbass AA, Nowoweiski SJ, Bernier D, Tarzwell R, Beutel ME (2014) Review of psychodynamic psychotherapy neuroimaging studies. Psychother Psychosom 83:142–147

Alberini CM, Travaglia A (2017) Infantile amnesia: a critical period of learning to learn and remember. J Neurosci 37:5783–5795

Andersen SM, Przybylinski E (2012) Experiments on transference in interpersonal relations: implications for treatment. Psychotherapy 49:370

Anderson M, Green C (2001) Suppressing unwanted memories by executive control. Nature 410:366–369

Anderson MC, Hanslmayr S (2014) Neural mechanisms of motivated forgetting. Trends Cogn Sci 18:279–292

Anderson MC, Ochsner KN, Kuhl B, Cooper J, Robertson E, Gabrieli SW, Glover GH, Gabrieli JD (2004) Neural systems underlying the suppression of unwanted memories. Science 303:232–235

Anzieu D (1996) Das Haut-Ich. Suhrkamp, Frankfurt a. M

Arbeitskreis OPD (Hrsg) (2006) Operationalisierte Psychodynamische Diagnostik. Hans Huber, Bern

Baddeley A (2003) Working memory: looking back and looking forward. Nat Rev Neurosci 4:829–839

Badre D, Nee DE (2018) Frontal cortex and the hierarchical control of behavior. Trends Cogn Sci 22:170–188

Bazan A (2017) Alpha synchronization as a brain model for unconscious defense: an overview of the work of Howard Shevrin and his team. Int J Psychoanal 98:1443–1473

Bion WR (1962) A theory of thinking. Int J Psychoanal 43:306–310

Blass RB, Carmeli Z (2007) The case against neuropsychoanalysis: on fallacies underlying psychoanalysis' latest scientific trend and its negative impact on psychoanalytic discourse. Int J Psychoanal 88:19–40

Brenner C (1973) An elementary textbook of psychoanalysis (revised ed). International Universities Press, New York. (Original work published 1955)

Buchheim A, Viviani R, Kessler H, Kächele H, Cierpka M, Roth G, George C, Kernberg OF, Bruns G, Taubner S (2012) Changes in prefrontal-limbic function in major depression after 15 months of long-term psychotherapy. PLoS ONE 7:e33745

Cahill L, McGaugh JL (1998) Mechanisms of emotional arousal and lasting declarative memory. Trends Neurosci 21:294–299

Carhart-Harris RL, Friston KJ (2010) The default-mode, ego-functions and free-energy: a neurobiological account of Freudian ideas. Brain 133:1265–1283

Chalmers DJ (1995) Facing up to the problem of consciousness. J Conscious Stud 2:200–219

Damasio A (2003) Feelings of emotion and the self. Ann N Y Acad Sci 1001:253–261

Davidson R, McEwen BS (2012) Social influences on neuroplasticity: stress and interventions to promote well-being. Nat Neurosci 15:689–695

de Waal FB, Preston SD (2017) Mammalian empathy: behavioural manifestations and neural basis. Nat Rev Neurosci 18(2017):72

Dennett DC (2018) Facing up to the hard question of consciousness. Philos Trans R Soc B Biol Sci 373:20170342

Dixon NF, Lear TE (1963) Electroencephalograph correlates of threshold regulation. Nature 198:870–872

Dollard J, Miller NE (1950) Personality and psychotherapy. McGraw-Hill, New York

Dresler M, Wehrle R, Spoormaker VI, Steiger A, Holsboer F, Czisch M, Hobson AJ (2015) Neural correlates of insight in dreaming and psychosis. Sleep Med Rev 20:92–99

Erdelyi MH (1985) Psychoanalysis: Freud's cognitive psychology. W. H. Freeman and Company, New York

Erdelyi MH (2006) The unified theory of repression. Behav Brain Sci 29:499

Fellner MC, Waldhauser GT, Axmacher N (2020) Tracking selective rehearsal and active inhibition of memory traces in directed forgetting. Curr Biol 30:2638–2644

Fischmann T, Russ MO, Leuzinger-Bohleber M (2013) Trauma, dream, and psychic change in psychoanalyses: a dialog between psychoanalysis and the neurosciences. Front Hum Neurosci 7:877

Fonagy P, Gergely G, Target M (2007) The parent–infant dyad and the construction of the subjective self. J Child Psychol Psychiatry 48:288–328

Fotopoulou A, Tsakiris M (2017) Mentalizing homeostasis: the social origins of interoceptive inference. Neuropsychoanalysis 19:3–28

Freud S (1900) Die Traumdeutung. GW II/III

Freud S (1914a) Erinnern, Wiederholen, Durcharbeiten. GW X, 126–136

Freud S (1914b) On Narcissism. SE XIV, 67–102

Freud S (1915a) Triebe und Triebschicksale. GW X, 210–232

Freud S (1915b) Die Verdrängung. GW X, 248–261

Freud S (1915c) Das Unbewusste. GW X, 264–303

Freud S (1923) Das Ich und das Es. GW XIII, 237–289

Freud S (1950) Entwurf einer Psychologie. GW Nachtragsband, 375–386 (Original work 1895)

Frith CD, Frith U (2012) Mechanisms of social cognition. Annu Rev Psychol 63:287–313

Gallese V (2009) Mirror neurons, embodied simulation, and the neural basis of social identification. Psychoanal Dialogues 19:519–536

Geiselman RE, Bjork RA, Fishman DL (1983) Disrupted retrieval in directed forgetting: a link with posthypnotic amnesia. J Exp Psychol Gen 112:58–72

References

Gerber AJ, Peterson BS (2006) Measuring transference phenomena with fMRI. J Am Psychoanal Assoc 54:1319–1325

Grünbaum A (1985) The foundations of psychoanalysis, A philosophical critique. University of California Press, Berkeley

Hassabis D, Kumaran D, Summerfield C, Botvinick M (2017) Neuroscience-inspired artificial intelligence. Neuron 95:245–258

Hobson JA (2009) REM sleep and dreaming: towards a theory of protoconsciousness. Nat Rev Neurosci 10:803–813

Horikawa T, Tamaki M, Miyawaki Y, Kamitani Y (2013) Neural decoding of visual imagery during sleep. Science 340:639–642

Jung CG, Ricklin F (1906) Experimentelle Untersuchungen über Assoziationen Gesunder. In: Jung CG (ed) Diagnostische Assoziationsstudien, Bd I. Johann Ambrosius Barth, Leipzig, 7–145

Kandel ER (1999) Biology and the future of psychoanalysis: a new intellectual framework for psychiatry revisited. Am J Psychiatry 156:505–524

Kaplan-Solms K, Solms M (2000) Clinical studies in neuro-psychoanalysis: introduction to a depth neuropsychology. Karnac Books, London

Kohut H (1983) Narzißmus: Eine Theorie der psychoanalytischen Behandlung narzißtischer Persönlichkeitsstörungen. Suhrkamp, Frankfurt a. M

Koukkou M, Lehmann D (1998) Ein systemtheoretisch orientiertes Modell der Funktionen des menschlichen Gehirns und die Ontogenese des Verhaltens – Eine Synthese von Theorien und Daten. In: Koukkou M, Leuzinger-Bohleber M, Mertens W (ed) Erinnerung von Wirklichkeiten – Psychoanalyse und Neurowissenschaften im Dialog, Bd 1. Verlag Internationale Psychoanalyse, Stuttgart, 287–415

Kragh U (1960) The defense mechanism test: a new method for diagnosis and personnel selection. J Appl Psychol 44:303–309

Kringelbach ML, Rolls ET (2004) The functional neuroanatomy of the human orbitofrontal cortex: evidence from neuroimaging and neuropsychology. Prog Neurobiol 72:341–372

Lacan J (1948) Das Spiegelstadium als Bildner der Ichfunktion, wie sie uns in der psychoanalytischen Erfahrung erscheint. In: Lacan J (ed) Schriften I. Quadriga, Weinheim, 61–70

Lane RD, Ryan L, Nadel L, Greenberg L (2015) Memory reconsolidation, emotional arousal, and the process of change in psychotherapy: new insights from brain science. Behav Brain Sci 38:e1

Leichsenring F, Leweke F, Klein S, Steinert C (2015) The empirical status of psychodynamic psychotherapy – an update: Bambi's alive and kicking. Psychother Psychosom 84:129–148

Levinger G, Clark J (1961) Emotional factors in the forgetting of word associations. J Abnorm Psychol Soc Psychol 62:99–105

Loftus EF (2017) Eavesdropping on memory. Annu Rev Psychol 68:1–18

Lorenzer A (2002) Die Sprache, der Sinn, das Unbewußte: Psychoanalytisches Grundverständnis und Neurowissenschaften. Klett-Cotta, Stuttgart

Mary A, Dayan J, Leone G, Postel C, Fraisse F, Malle C, Vallée T, Klein-Peschanski C, Viader F, de la Sayette V, Peschanski D, Eustache F, Gagnepain P (2020) Resilience after trauma: the role of memory suppression. Science 367:eaay8477

Mendelsohn A, Chalamish Y, Solomonovich A, Dudai Y (2008) Mesmerizing memories: brain substrates of episodic memory suppression in posthypnotic amnesia. Neuron 57:159–170

Moscovitch M, Cabeza R, Winocur G, Nadel L (2016) Episodic memory and beyond: the hippocampus and neocortex in transformation. Annu Rev Psychol 67:105–134

Nader K, Hardt O (2009) A single standard for memory: the case for reconsolidation. Nat Rev Neurosci 10:224–234

Panksepp J (1998) Affective neuroscience – the foundations of human and animal emotions. Oxford University Press, New York

Panksepp J (2011) Cross-species affective neuroscience decoding of the primal affective experiences of humans and related animals. PLoS ONE 6:e21236

Papoušek H, Papoušek M (1987) Intuitive parenting: a dialectic counterpart to the infant's integrative competence. In: Osofsky JD (ed) Wiley series on personality processes. Handbook of infant development. Wiley, 669–720

Parsons CE, Young KS, Murray L, Stein A, Kringelbach ML (2010) The functional neuroanatomy of the evolving parent–infant relationship. Prog Neurobiol 91:220–241

Rapaport D (1971) Emotions and memory (5th ed). International Universities Press, New York (Original work published 1946)

Rilling JK, Young LJ (2014) The biology of mammalian parenting and its effect on offspring social development. Science 345:771–776

Sakai J (2020) Core concept: how synaptic pruning shapes neural wiring during development and possibly, in disease. Proc Natl Acad Sci 117:16096–16099

Salvador A, Berkovitch L, Vinckier F, Cohen L, Naccache L, Dehaene S, Gaillard R (2018) Unconscious memory suppression. Cognition 180:191–199

Schmeing JB, Kehyayan A, Kessler H, Do Lam AT, Fell J, Schmidt AC, Axmacher N (2013) Can the neural basis of repression be studied in the MRI scanner? New insights from two free association paradigms. PLoS ONE 8(4):e62358

Shevrin H, Snodgrass M, Brakel LA, Kushwaha R, Kalaida NL, Bazan A (2013) Subliminal unconscious conflict alpha power inhibits supraliminal conscious symptom experience. Front Hum Neurosci 7:544

Solms M (2000) Dreaming and REM sleep are controlled by different brain mechanisms. Behav Brain Sci 23:843–850

Solms M (2018) The neurobiological underpinnings of psychoanalytic theory and therapy. Front Behav Neurosci 12:294

Solms M (2019) The hard problem of consciousness and the free energy principle. Front Psychol 9:2714

Solms M (2020) New project for a scientific psychology: general scheme. Neuropsychoanalysis 22:5–35

Solms M, Turnbull OH (2011) What is neuropsychoanalysis? Neuropsychoanalysis 13:133–145

Squire LR (2004) Memory systems of the brain: a brief history and current perspective. Neurobiol Learn Mem 82:171–177

Stern DN (2010) Die Lebenserfahrung des Säuglings. Klett-Cotta, Stuttgart

Storck T (2020) 100 Jahre Rezeption des Todestriebkonzepts. Psyche 74:831–867

Waldhauser GT, Johansson M, Hanslmayr S (2012) Alpha/Beta oscillations indicate inhibition of interfering visual memories. J Neurosci 32:1953–1961

References

Waldhauser GT, Bäuml K-HT, Hanslmayr S (2015) Brain oscillations mediate successful suppression of unwanted memories. Cereb Cortex 25:4180–4190

Waldhauser GT, Dahl MJ, Ruf-Leuschner M, Müller-Bamouh V, Schauer M, Axmacher N, Elbert T, Hanslmayr S (2018) The neural dynamics of deficient memory control in heavily traumatized refugees. Sci Rep 8:13132

Warsitz RP, Küchenhoff J (2015) Psychoanalyse als Erkenntnistheorie – psychoanalytische Erkenntnisverfahren. Kohlhammer, Stuttgart

Weiner B (1968) Motivated forgetting and the study of repression. J Pers 36:213–234

Winnicott DW (1960) The theory of the parent–infant relationship. In: Winnicott DW (ed) (1965). The maturational processes and the facilitating environment: studies in the theory of emotional development. Hogarth, London, 37–55

Wiswede D, Taubner S, Buchheim A, Münte TF, Stasch M, Cierpka M, Kächele H, Roth G, Erhard P, Kessler H (2014) Tracking functional brain changes in patients with depression under psychodynamic psychotherapy using individualized stimuli. PLoS ONE 9:e109037

Yovell Y, Solms M, Fotopoulou A (2015) The case for neuropsychoanalysis: why a dialogue with neuroscience is necessary but not sufficient for psychoanalysis. Int J Psychoanal 96:1515–1553

Zeller AF (1950) An experimental analogue of repression. II. The effect of individual failure and success on memory measured by relearning. J Exp Psychol 40:411–422

Printed in the USA
CPSIA information can be obtained
at www.ICGtesting.com
LVHW090539060124
768272LV00010B/430